HOW DC MOTOR WORKS?

Fundamental Concept & Mathematical Representation

Dr. Jignesh A. Makwana
Ph.D. Electrical, Indian Institute of Technology Roorkee, India

HOW DC MOTOR WORKS?

Copyright © 2018, 2019 by Dr. Jignesh A. Makwana.

All rights reserved.

All rights reserved.

No part of this book may be reproduced in any form, by photostat, microfilm, xerography, or any other means, or incorporated into any information retrieval system, electronic or mechanical, without the written permission of the author.

Edition: First Edition (2018-2019)

Imprint: Independently published

For information contact :

www.rhyni.com

jignesh.makwana@rhyni.com

Tech Skills & Fundamentals

I dedicate this book to my daughter Rhythm, for allowing me the time to write it. I carry your mesmerizing smile and whispering words throughout the writing of the book. I may not be able to carry you in my arms by the time, but I will always carry you in my heart.

I dedicate this book to my mother, my father, my grandmother, my sister, other family members, my friends, my teachers, my guides, and to everyone who helped shape me into the person that I am.

PREFACE

The primary objective of this book is to explain "How DC Motor works ?" in a simplified way. This book is designed to help students in understanding fundamental concepts and mathematical derivation. I focus on new learning perspectives for students of schools & colleges. "Easy to understand" and "Every bit to understand" are two basic blocks of this book.

This book is designed for the audience how curious about "How electric motor works?", wants to improve fundamentals of DC Motor, students ages from school &college, first-year engineering students& students of electrical engineering &related branch.

What you'll learn from this book?

- Will have answer of "Why motor starts rotating when supply by voltage source?".
- Will understand role of back-emf in energy conversion process.
- Will understand "How to derive equations of motor speed?".
- Will understand torque-speed relation and its graphical representation
- Will understand speed control methods of DC Motor.

The author is grateful to Dr. Dinesh Kumar, Ph.D., University of Coimbra, Portugal, for checking & proofreading. The author is grateful to Dean Prof. R.B Jadeja & Prof. Sarang Pande for their motivation and support. The author is also grateful to all faculty members of the electrical engineering department for their moral support.

The author welcomes any constructive criticism of the book and will be grateful for any appraisal by the readers.

CONTENTS

Chapter 1: Fundamental Concept………………………..... 13
- ✓ Fundamental of Electrical Motors…………………... 14
- ✓ How electric current develops mechanical force & torque……………………………………………...…... 16
- ✓ How to supply electric current to the rotating coil…….. 24
- ✓ How to maintain the direction of current flowing through conductors…………………………………….. 26
- ✓ Important role of Faraday's law and Lenz's law in energy conversion (Back emf)…………………………. 30

Chapter 2: Mathematical Representation………………….. 33
- ✓ How to produces torque and speed of desired quantity.. 34
- ✓ Graphical Representation of Torque-Speed equation…. 38

Chapter 3: Speed Control of DC Motor…………………….. 39
- ✓ Effect of Supply Voltage………………………………. 40
- ✓ Effect of adding resistance to the armature winding…... 40
- ✓ Effect of adding resistance to the field winding……….. 41
- ✓ Summary……………………………………………….. 43

Chapter 4: More About DC motor…………………………..... 44
- ✓ Controversy effect of field flux on motor torque………. 45
- ✓ Equation of force &torque……………………………... 46
- ✓ Shape and equation of back emf………………………. 47
- ✓ Average value of back emf…………………………….. 49

- ✓ Power transfer... 49
- ✓ DC series motor... 50
- ✓ Shaft Torque... 55
- ✓ Applications and limitations............................... 55

Appendix:... 56
- ✓ Faraday's law.. 57
- ✓ Lenz's law.. 57
- ✓ Screw rule.. 57
- ✓ Ohm's law.. 58

HOW DC MOTOR WORKS?

Chapter 1: Fundamental Concept

This chapter explains how mechanical force produces by electric current and the arrangement required to produce rotary motion.

Fundamental of Electrical Motors:

Electrical motor converts electrical energy into mechanical energy. The main parameters of electrical systems are voltage, current, and resistance. While parameters of the mechanical system are speed, torque (force), and friction. Electrical Power is a product of voltage and current. That means if we consider voltage source is constant, power is proportional to current, and thus current demand increases with power. Whereas mechanical power is a product of speed and torque. That means for the same power output requirement, heavier load can be driven at low speed, or higher speed can be achieved at lighter load.

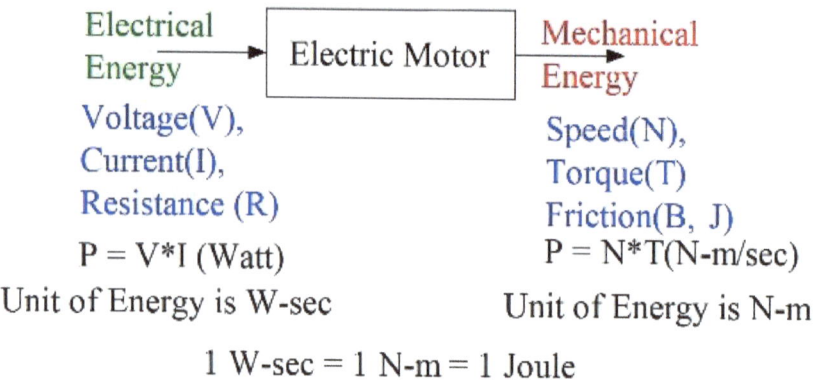

Fig. 1 Energy Conversion

Consider basic electric circuits having voltage source and constant resistance, then according to Ohm's law (V = IR), current becomes directly proportional to the voltage. More voltage results in more current. Like in a mechanical system more torque results more speed (angular velocity or rotation) considering constant friction.

Fig. 2 Bearing of electric motor

Electric Motor is a device where the shaft rotates when a motor is supplied by voltage (AC or DC). It is assumed here that everyone knows about the fact that bearing attached with shaft allows mechanical structure to rotate freely (least friction) and so there is no discussion about bearing in further discussion. Thus, understanding the working of an electric motor means understanding why the shaft rotates when a motor is supplied by the voltage. The same thing is represented as block diagram view in Fig.3. As already discussed, when voltage produces current, and torque develops speed, an electric motor can be further streamlined as shown in Fig.4. Here supplying voltage results in current, current results in the development of torque, and torque results in speed. Thus, understanding how an electric motor works, is to understand how electrical current develops mechanical torque. It's required to understand few basic principles of electromagnetic and required mechanical arrangements which are explained in the next section.

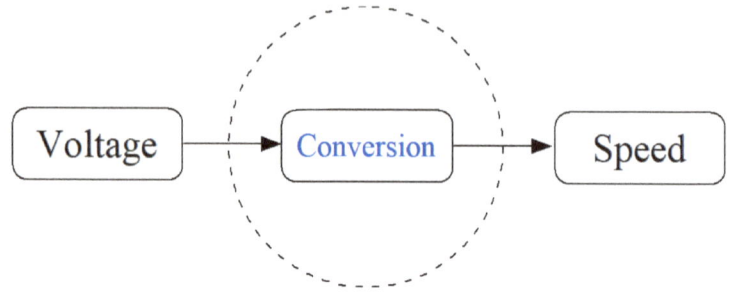

Fig. 3 Representation of electric motor as one block

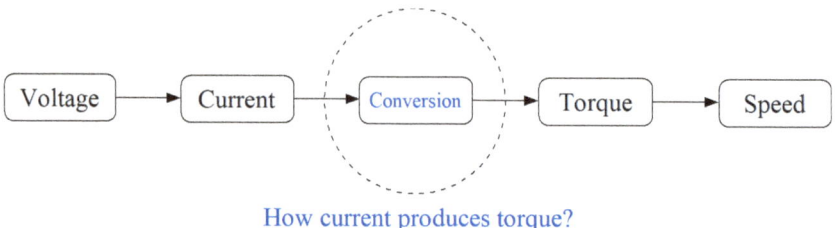

Fig. 4 Representation of electric motor in electrical & mechanical parameters

How electric current develops mechanical force & torque.

Before understanding torque development from the current, let's discuss the things we all aware of, yes, it's about permanent magnet. Two magnets are attracted or repulsed to each other depends upon polarity, the same polarity repulse and opposite polarity attracts as demonstrated in fig.5. That means natural mechanical force develops between two magnets. It should be noted that a magnet does not possess or store energy, but when it attracts another magnet, energy is stored and we need to apply the same amount of energy to detach the same magnets.

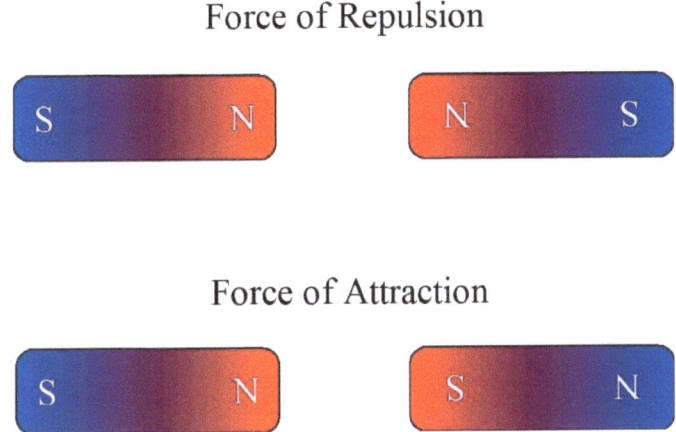

Fig. 5 Force produced in permanent magnet

Consider the case where one magnet is fixed and the other is free to move as shown in Fig.6. Here force of repulsion experienced by the magnet will result in its displacement and velocity. Here mechanical force is produced due to the interaction of two flux or say magnetic field.

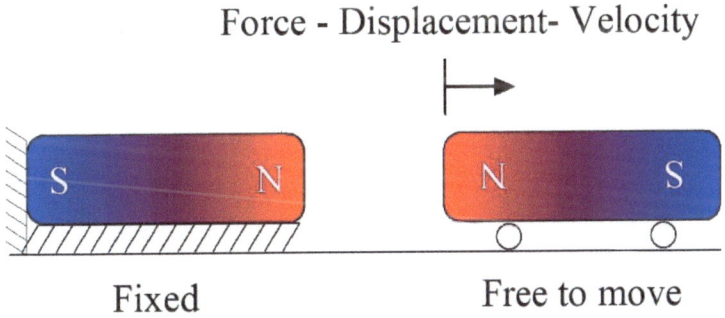

Fig. 6 Force results in velocity and displacement

Good thing is that we can produce the same magnetic flux with the help of electric current and coil too. When electric current flows through the conductor, it produces magnetic flux around its periphery. The direction of current and flux produced are shown in Fig. 7 & 8. When current flows inward, magnetic field produce in anti-clockwise direction and when current flows outward, magnetic field produce in clockwise direction. To remember the direction of flux produced, crew rule and left-hand rule are used which are explained in the appendix. If we consider the current carrying coil, it will produce a resultant flux of the same pattern as a permanent magnet as shown in Fig. 9. That means any current-carrying coil behaves as a magnet. In general, the coil is wound on iron instead of air-core as it allows low reluctance path than air and thus produces more concentrated flux. Such device which produces such magnets by using coil and core(iron) called "Electromagnet".

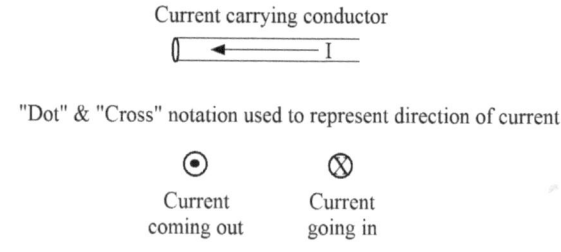

Fig. 7 Notation of direction of current flowing through conductor

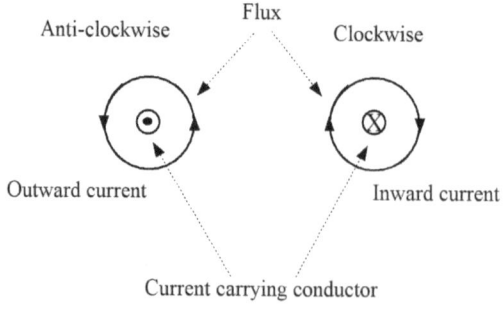

Fig. 8 Direction of flux produces in current-carrying conductor

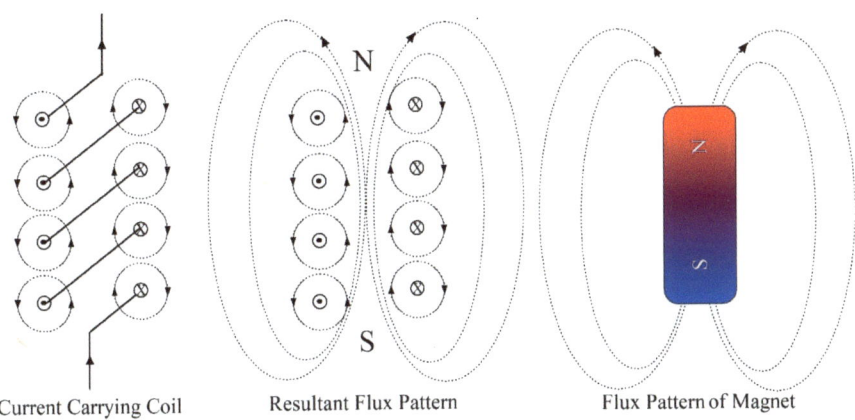

Current Carrying Coil   Resultant Flux Pattern   Flux Pattern of Magnet

Fig. 9 Pattern of flux produces on current-carrying coil

Consider a case shown in Fig. 10. Here one is a fixed electromagnet and the other is a permanent magnet that is free to move. Now we also have controllability as when we switch-on the supply, flux will be developed which produces the force as shown in Fig.10. Interesting thing is that now we have the answer of "How current produces torque(force)" discussed in Fig.4. Yes, it's magnet & magnetic flux. Conversion of electrical energy to mechanical force/torque happens due to the medium of magnetic flux. Thus, the process of the electric motor shown in Fig. 4 can be represented more precisely as shown in Fig. 11. Supply voltage produces current, current produces flux, interaction of flux produces force(torque) and force(torque) produces displacement(rotation) and velocity(speed).

Up to now, we have discussed producing force and linear displacement. The basic force-producing principle we studied is force produced between two magnets or say between electromagnet and magnet. More specifically said, force is produced by the **interaction of two fluxes (magnet)**.

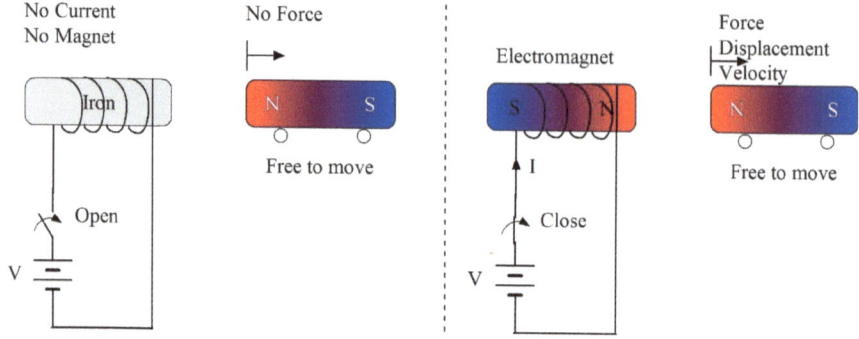

Fig 10 Force produces with the help of current (electromagnets)

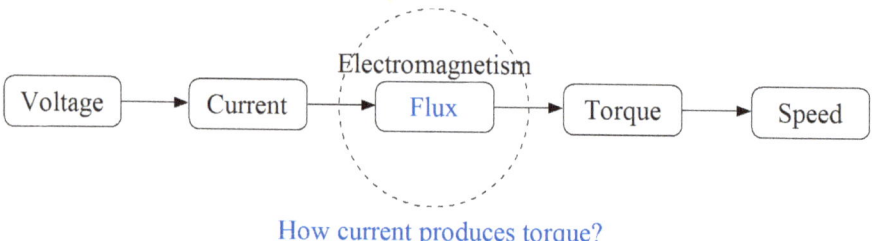

How current produces torque?

Fig. 11 Representation of electric motor as electrical, mechanical & magnetic parameter.

Let's understand which arrangement is required to have rotary motion. We are going to utilize the same principle where the interaction of two fluxes produces the force. Let's discuss further.

Consider an arrangement shown in Fig. 12. It is a case when the current carrying conductor is placed in a magnetic field. Here one magnetic field is developed due to permanent magnets and the second will be developed due to the current-carrying conductor as shown in Fig. Let's discuss the interaction of two fluxes. Magnetic flux line flows from N pole to S pole and conductor where current flowing outward produces surrounded flux in anti-clockwise direction. When these two fluxes interact with each other, the total flux will be additive on the left side of the conductor, while subtractive on the right side. It results in more concentration of magnetic flux on the left side and less on the right side. This will develop mechanical force on the conductor from left to right. And case will be opposite when changing the direction of current flowing through the conductor as shown. **So, it is understood that when we put a current-carrying conductor in a magnetic field, it experiences a force.**

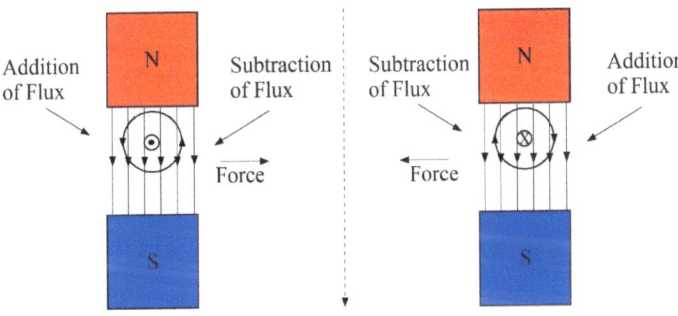

Fig. 12 Force on current-carrying conductors placed in magnetic field

Now consider a coil having two conductors as shown in Fig. 13(a). Note that we are considering only two sides of the coil which are placed under the magnetic field. The other two sides will remain outside the magnetic field and remain important only to provide the closed path and will not contribute to developing torque. The same coil can be represented as shown in Fig.13(b) where current flowing to both the conductors will be opposite to each other. For the period, assume that we have an arrangement to supply electric current to the rotating coil. However later on further discussion, we will see how to do that. Thus, force developed on both conductors will also be opposite to each other. Here idea is to utilize these developed forces to produce rotary motion. Consider an arrangement shown in Fig. 14 where a rotor is made up of cast iron and has slots for placement of conductor. Bearing is used to allow free rotation of structure as it provides the least friction. When the coil is supplied by the current in the direction shown, force will be developed and the rotor will start rotating in a clockwise direction as shown in Fig. 15. Everything is fine till 90° of rotation, but after that same force opposes the direction of rotation. Consider a centerline between N and S pole as a magnetic neutral axis. So, to have a continuous rotation in the same direction it is required to produce force in the same direction. It requires changing the direction of current flowing through the conductor when it crosses the magnetic neutral axis. In other words, the current flowing through the conductors above the magnetic neutral axis should remain in the same direction throughout to produce the rotatory motion. Thus, conductors under N poles and S poles needs to maintain the same current in opposite direction. It requires an arrangement that changes the direction of current flowing through the conductor when it crosses the magnetic neutral axis. Let's understand how we do this.

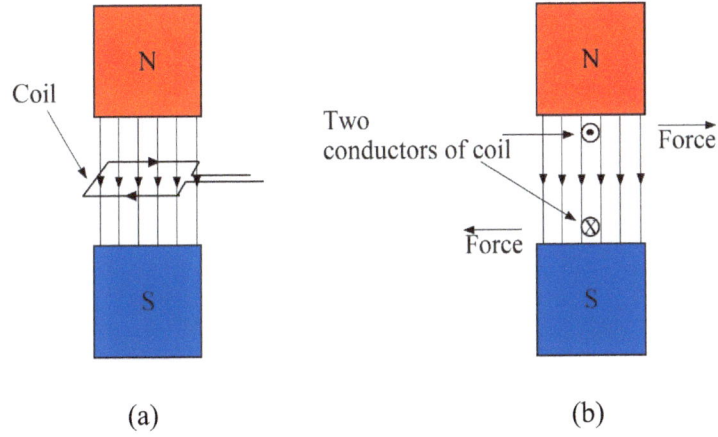

Fig. 13 Current carrying coil in magnetic field

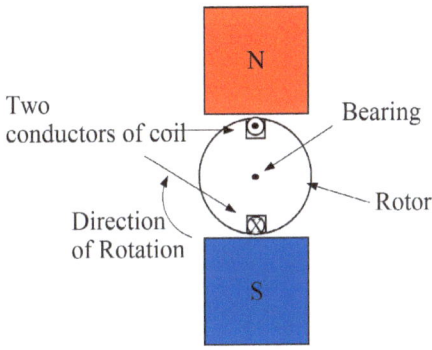

Fig. 14 Placement of conductors on rotor

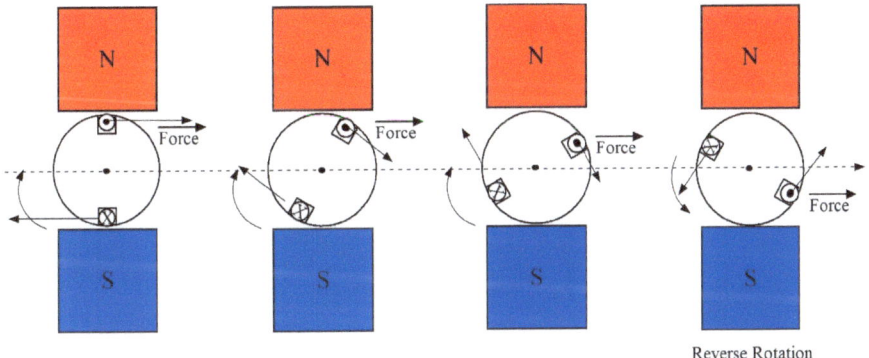

Fig. 15 Direction of force and rotation when supply DC to conductors

How to supply electric current to the rotating coil

First understand how can we supply current to rotating conductors. The very basic arrangement used in general is shown in Fig. 16. Here two fixed conductive rings are used to supply the current to the coil. This kind of arrangement is shown in fair & amusement parks to supply toy train or bumping cars shown in Fig.17. Here overhead wire will always remain in touch with conductive ceiling mesh even when the car is moving. Same way, the conductors of the coil in Fig. 16 remains always in contact with the fixed coil even when rotating. But when conductors are not thick enough to maintain contact, we require extra thick rods or rigid arrangement at the end of the coil. It's called slip-ring and brush arrangement shown Fig.18. Here each end of the conductor is in fixed contact with a conductive ring called a slip-ring, and this ring is also rotating with the rotor. Brushes are fixed and maintain contact with rotating conductive rings. Thus, this arrangement is used to supply current to the rotating coil. But still requirement of maintaining the direction of current flowing through conductors positioned above and below the magnetic neutral line is not fulfilled.

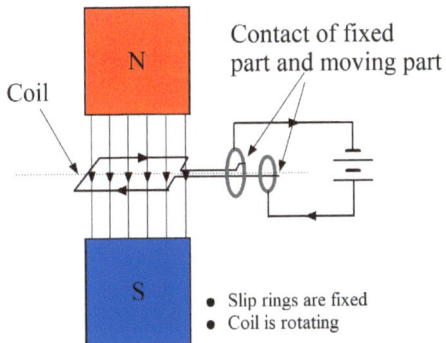

Fig. 16 Arrangement of conductive rings which supply current to rotating conductors

Fig. 17 Electric supply to moving object (bumping cars) from fixed voltage source

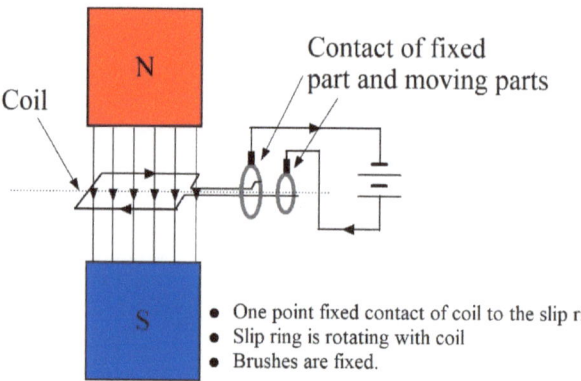

Fig. 18 Arrangement of brushes and slip-rings to supply current to rotating conductors

How to maintain the direction of current flowing through conductors

Let's consider the arrangement shown in Fig 19(a). Here conductive ring is divided into two segments for two conductors. Two segments are isolated from each other. Here we can see air as an insulating medium between two segments of conductive ring, but in actual practice, we can have insulating material like mica. The placement of brushes plays an important role in maintaining desired current direction. Fig. 19(b) shows a more clear view of brush placement for maintaining desired current direction to produces rotary motion. The rotation of the rotor from 0° position to 120° is shown in Fig.20. Observe change in current direction when conductor crosses the magnetic neutral axis (i.e 90°) as seen in Fig.20(d) and Fig.20(e). Thus, it is required to have segment arrangement instead of coil, and placement of brushes should be on magnetic neutral axis to develop torque in one direction. This segmental arrangement is called a **commutator**. The main role of the commutator is to maintain the direction of current flowing through the conductor.

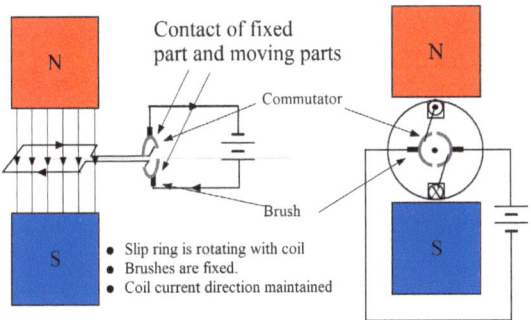

Fig. 19 Arrangement of commutator to maintain current direction

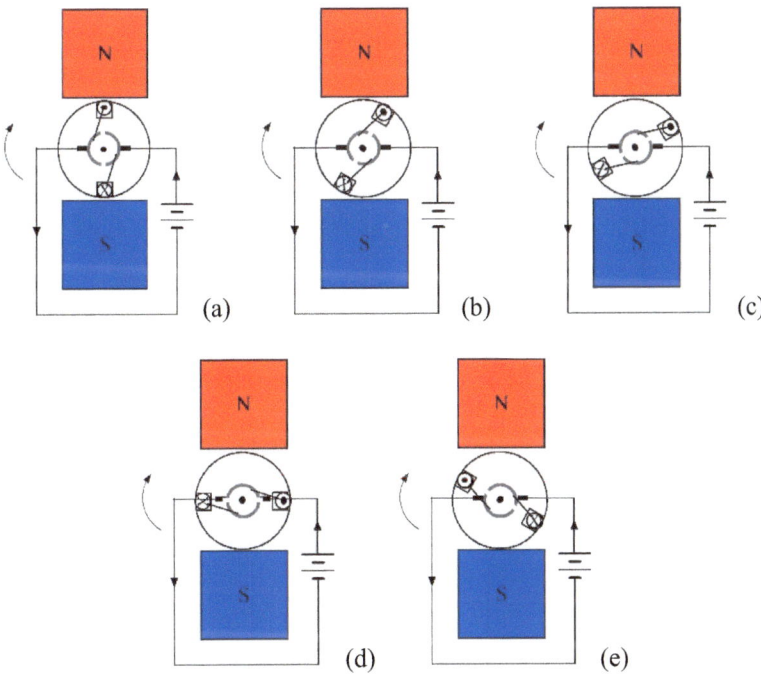

Fig. 20 Demonstration of motor rotation with commutator and proper placement of brushes (a) θ = 0° (b) θ = 30°(c) θ = 60° (d) θ = 90°(e) θ =120°

This is how DC motor works. Supplying voltage through brushes and commutator arrangement maintains the direction of current flowing through conductors. Do not forget the importance of brushes placement here. Magnetic fields developed by permanent magnets will interact with flux produce by current-carrying conductors and develop force. Thus, it is said that the current-carrying conductor placed in the magnetic field results in force and so rotation of the rotor in the DC motor. As permanent magnets are used to produce flux, these types of motors are named **Permanent Magnet DC Mot**or say **PMDC**. These types of motors are used in low-power applications like electric toys and domestic applications.

Observe the construction of DC Motor shown in Fig. 21. Here electromagnet is used instead of the permanent magnet to produce the flux. Coils are wound around the iron **pole** (also called **pole shoe**) such that one produces an N pole and the other produces an S pole. These coils are connected in series and form one winding called **filed winding**. Structure provides support to poles and complete close path of a magnetic circuit called **Yoke**. It is made up of magnetic material like iron to provide less reluctance to the flux as compared to air. Complete structure of Yoke, Poles, and Field coil called **stator** as no rotating parts are there. Rotting assembly including core, conductors, and commutator called **rotor** or **armature**. More numbers of conductors and thus more numbers of commutator segments are used in actual motor construction to produce maximum possible torque. These conductors of coils form a winding called **armature winding**. These types of DC Motor are called **Separately Excited DC Motor** as separate supply is required by the field winding. But there is no reason to supply through a separate voltage source in the case of a DC motor. Thus, these types of DC motor are not in general practice, but it is related to DC Generator where separately excited DC Generators are widely known. In the case of DC Motor same supply is used for both armature winding and field winding by connecting both in parallel and it's called a **Shunt DC Motor**. Now if we list the main part that construct DC motor, they are armature, poles, yoke, armature winding, field winding, commutator, and brushes.

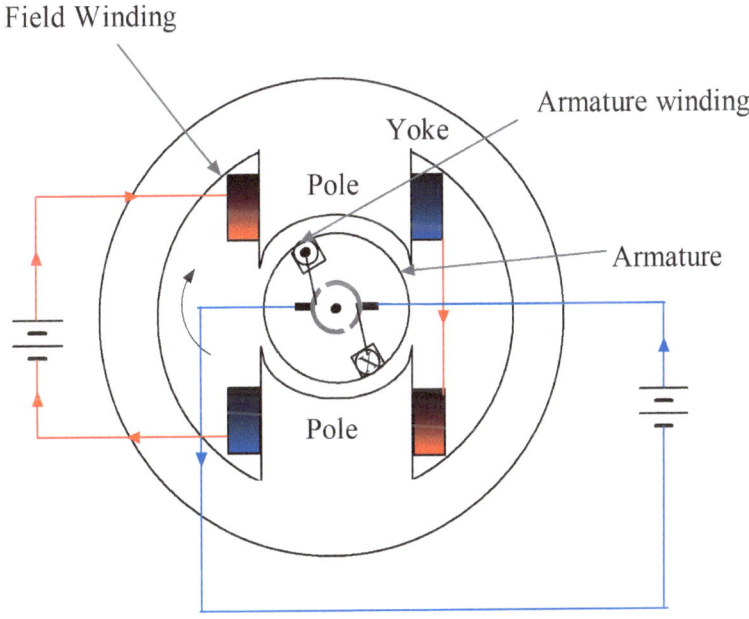

Fig. 21 Basic Parts of DC Motor

Important role of Faraday's law and Lenz's law in energy conversion (Back emf)

We understood that how supplying voltage results in torque and speed. Now think about what's happen when we increase the mechanical load to the motor. Yes, as we put more mechanical load, the motor should develop more torque to maintain the desired speed, which means armature winding will draw more current from the voltage source. Yes, it happens automatically and can explained with help of Faraday's law and Lenz's law.

As we know that Faraday's law is about generating electricity or say emf (electromotive force or voltage) with the help of coil and magnet. When flux linked with the coil or conductor changes, an emf will be induced in it. And direction of induced emf is given by Lenz's law. Emf will induce in such direction so that it opposes the quantity which is responsible for its production. Yes, its law of opposite sign, represented by negative sign in equation of induced emf $e = - d\phi/dt$.

Now consider DC motor in operation. The rotor is rotating due to supply voltage. Here flux produces by the magnetic poles linked with the conductors. So, when the motor rotates, fluxed linked with armature conductors will change and so emf will induced in conductor. Thus, when the motor is running, an induced emf will always there across armature winding. According to Lenz's law, the direction of this induced emf will be such that it opposes the supply voltage or supply current. Thus, voltage appeared across armature winding will always difference of supply voltage and induced emf. Here supply voltage is constant when induced emf is proportional to the rate of change of flux and so proportional to the motor speed. So more speed results in less voltage appeared across armature winding results in less current.

Consider a case when the motor is running with constant speed at no load. As we increase the mechanical load to the motor, speed tries to get reduce. As speed drops, induced emf also reduces results increase in net voltage appeared across armature winding. Thus, it will draw more current in response to an increase in mechanical load. Here induced emf is called **back emf** as it opposes the supply voltage.

This is how Faraday's law and Lenz's law play an important role in energy conversion. Thus, without understanding the concept of **back emf**, the process of energy conversion will never be fully understood.

Up to now, we understood the fundamental working principle of how DC motor works and the role of back emf. Do you want to know furthermore about DC motors? Let's understand how can we produce desired torque and desired speed of the rotor. Yes, we need to take the help of mathematic for this, which will give us an idea about how much current or voltage we required to produce the desire amount of torque and speed.

Chapter 2: Mathematical Representation

This chapter explains about mathematical relation among speed, torque, and other variables like supply voltage, armature resistance, field resistance, armature current & field current.

How to produces torque and speed of desired quantity?

Now we know that supplying voltage to the armature winding will result in rotation of the motor. Let's discuss how mathematically we can represent the effect we discussed. We are going to derive an equation that gives an idea about speed when the motor is supplied by the voltage of finite value. That means on the left-hand side of the equation we should have speed, and on the right-hand side of the equation, we need voltage and other variables of electrical and mechanical systems.

A simplified representation of a shunt DC motor in the form of an electric circuit is shown in Fig.22(a). Considering back emf Eb, the circuit can be re-draw as shown in Fig. 22(b). Let's first apply the knowledge of KVL & KCL (Kirchhoff's voltage & current laws) to the simplified electric circuit. Applying KVL to armature winding results in equation (1), applying KVL to field winding results in equation (2), and applying KCL to supply current results in equation (3).

$$V_{dc} = I_a R_a + E_b \tag{1}$$

$$V_{dc} = I_{sh} R_{sh} \tag{2}$$

$$I_{dc} = I_a + I_{sh} \tag{3}$$

If not forgotten, we required the equation where on the left-hand side we required speed and on the right-hand side, we required voltage and other variables. That means we required speed as a function of voltage. Mathematically it can be represented as equation (4) and can be said that speed is a function of voltage.

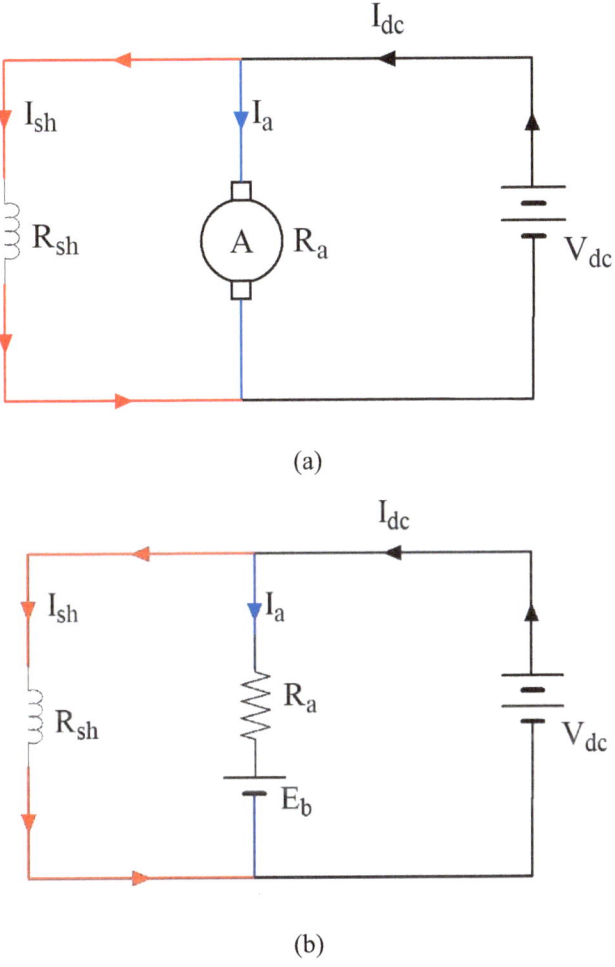

"A" – Represents Armature (rotor) supplied through brushes(fixed)
R_a is resistance of armature winding
R_{sh} is resistance of field winding
I_{dc} is current drawn from the DC supply
I_a is current drawn by the armature winding
I_{sh} is current drawn by the shunt field winding
V_{dc} is supply voltage
E_b is back emf

Fig. 22 Electric circuit diagram of shunt DC Motor

$$N = f(V_{dc}) \tag{4}$$

where N is the speed of the motor in rpm (revolution per minute) & V_{dc} is supply voltage. As far it seems speed N is not there in any of the equations we derived. But it's not true, don't forget that back emf is directly proportional to the speed. Thus, an equation for back emf can be written as,

$$E_b = K_b N \tag{5}$$

where K_b is constant. Constant K_b can be found easily if we know at least one relation between speed and back emf. (i.e if we know that E_b is 100V at the speed of 1000 rpm then K_b becomes 0.1). Putting value of K_b in equation (1), it becomes

$$V_{dc} = I_a R_a + K_b N \tag{6}$$

Now equation (6) has both speed and voltage. So, let's simplify the equation for the required format.

It becomes,

$$N = \frac{V_{dc} - I_a R_a}{K_b} \tag{7}$$

Equation (7) represents a relation between speed and supply voltage. It shows that an increase in supply voltage results increase in speed. Here R_a and K_b are constant but variable I_a does not looks good in this equation as it also depends on supply voltage & applied load torque. As per Ohm's law and also from equation (1), I_a will increase with voltage. However, increasing load torque will also increase I_a

momentarily. This is due to an increase in load results in a momentary reduction in speed and so a reduction in E_b. As E_b reduces, I_a will increase as referred by equation (1). And an increased current results in more torque development and thus increase the speed and back emf again. This is why I_a increases momentarily when increases load torque. So, let's try to eliminate I_a from equation (7). Let's represents torque developed by the motor by T_m in Newton-meter. As developed torque is proportional to the armature current, we can write,

$$T_m = K_t I_a \qquad (8)$$

where K_t is torque constant. Thus, after replacing the value of I_a, equation (7) becomes,

$$N = \frac{V_{dc} - \frac{T_m}{K_t} R_a}{K_b} \qquad (9)$$

Thus, equation (9) gives an idea about how motor speed changes with supply voltage and required torque. Here we have not discussed the load toque yet. If torque developed by the motor is T_m, let's consider the load torque T_L. As we have seen how motor current increases automatically in response to higher load torque, we can assume that the motor develops more torque when increases load torque T_L. So, neglecting friction and inertia of the rotor, torque developed by the motor T_m is equal to load torque T_L. Thus, if we say torque developed by the motor is very less, it means the motor is lightly loaded and vice versa. The current carrying capacity of the conductor decides the maximum torque of the motor. It also limits the maximum supply voltage.

Graphical Representation of Torque-Speed Equation.

Equation (9) reveals how speed varies with changes in voltage and load torque. Graphical representation of relation between speed and toque called speed-torque characteristics of the motor. It helps in understanding analyzing motor performance for different loads. Let's assume the motor is supply with rated voltage V_{dc} and plot variation of speed with torque. Speed at zero torque or say no load is called no-load speed. Putting $T_m = 0$ in equation (9), no-load speed becomes,

$$N_{nl} = \frac{V_{dc}}{K_b} \qquad (10)$$

And speed reduces as torque increases. Thus, no-load speed is the maximum possible speed of the motor. Variation of speed with torque is plotted in Fig. 23.

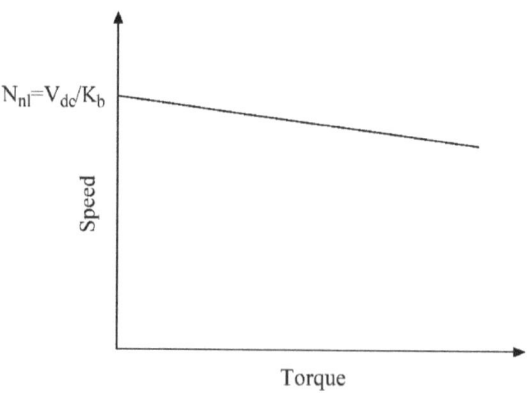

Fig. 23 Speed-torque characteristics

Chapter 3: Speed Control of DC Motor

Possible ways to control the speed of the DC motor are explained with mathematical & graphical support.

Effect of Supply Voltage

Equation (9) and (10) reveals that reduction in supply voltage results reduction of no-load speed. However, the effect of load current remains the same. The effect of voltage reduction on speed-torque characteristics is shown in Fig. 24. It shows motor speed can be reduced by reducing the supply voltage.

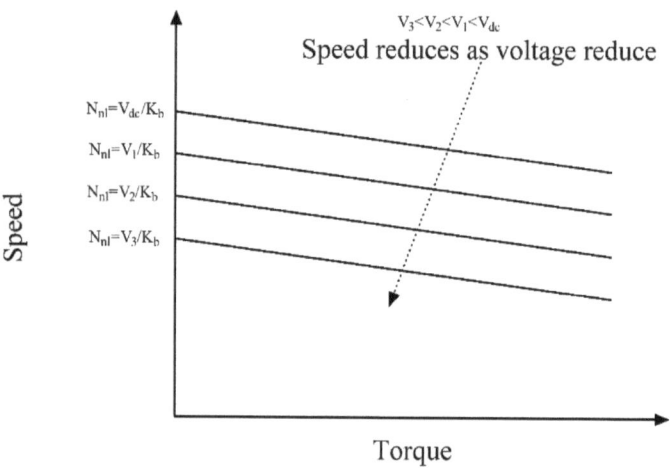

Fig. 24 Effect of voltage reduction on speed-torque characteristics

Effect of Adding Resistance to the Armature Winding

Equation (9) shows that armature resistance R_a also affects the motor speed. Even though the value of armature resistance is constant and very small, external resistance can be inserted in series with armature resistance to change the speed. An increase in resistance will results reduction in speed. Assuming constant V_{dc}, the effect of inserting external resistance on speed-torque characteristics is plotted in Fig.25. Here variation in resistance does not affect no-load speed or say maximum speed of the motor. Motor speed reduces with an increase in resistance.

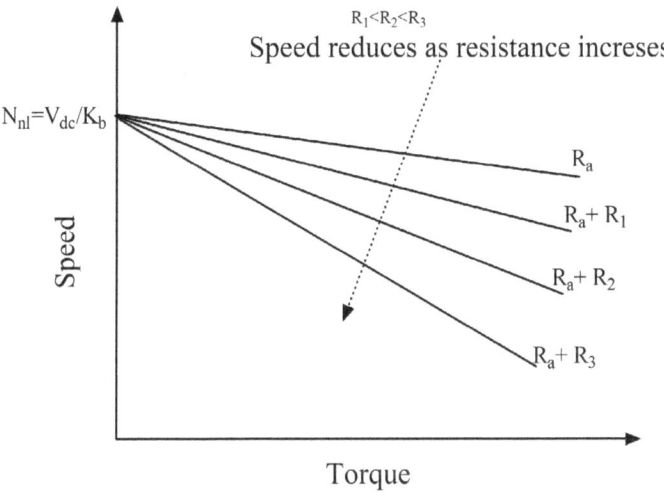

Fig. 25 Effect of adding external resistance to the armature winding

Effect of Adding Resistance to the Field Winding

Let's discussed what happens if we insert external resistance to field winding. In equation (5), Eb is proportional to motor speed as we had assumed constant field current (flux), however it also depends on field flux. Back emf reduces with reduction in field flux or say current. Thus, equation 5 can be re-write as,

$$E_b = K_b N \phi_f \qquad (11)$$

Where ϕ_f is field flux in weber. Equation of speed and no-load speed can be written as

$$N = \frac{V_{dc} - \frac{T_m}{K_t} R_a}{K_b \phi_f} \qquad (12)$$

$$N_{nl} = \frac{V_{dc}}{K_b \phi_f} \qquad (13)$$

Inserting external resistance to field winding results in a reduction in field current and field flux ϕ_f. Thus, it increases no-load speed. Here reduction in the flux is the result of adding external resistance to the field. Thus, we can say motor speed can be increased by adding external resistance to the field winding. Variation of speed-torque characteristics with external field resistance is shown in Fig. 26.

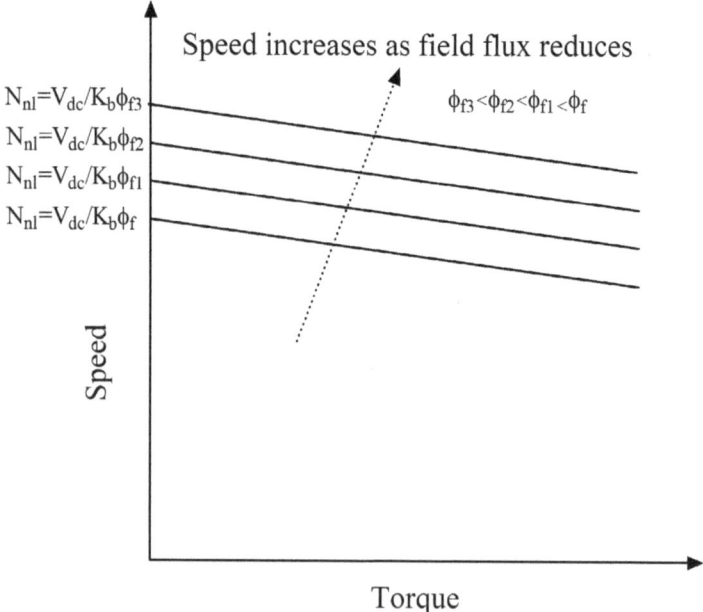

Fig. 26 Effect of adding resistance to the field winding

Thus, motor speed can be controlled by supply voltage or inserting external resistance in series with the armature winding. However, the second method is not preferable as inserting resistance in series with armature winding results in I²R losses and reduces motor efficiency.

Summary:

- ✓ Mechanical force produces due to the interaction of two magnetic fluxes.
- ✓ The basic operating principle of DC Motor is when the current carrying conductor is placed in a magnetic field it experiences a force.
- ✓ The commutator supplies current to the rotating coil and maintains its direction too.
- ✓ The placement of brushes plays an important role in maintaining the current direction.
- ✓ The main parts of DC Motor are armature, commutator, brushes, poles, yoke, armature winding, and field winding.
- ✓ Speed of DC motor governed by equation $N = \dfrac{V_{dc} - \dfrac{T_m}{K_t} R_a}{K_b}$
- ✓ Motor speed can be controlled by controlling supply voltage or inserting external resistance in series with the armature winding.
- ✓ Adding resistance to the field winding results increase in speed

Chapter 4: More About DC motor

Are you student belong to electrical engineering or related branch? Here is more detail about the DC motor for you. Let's explore.

Controversy Effect of Field Flux on Motor Torque Developed.

We have seen that reduction in field flux results in increased motor speed. However, as an electrical engineer, it is interesting to understand the controversial effect as motor torque also depends on field flux. Equation (8), $T_m = K_t I_a$ was derived by assuming constant field flux. As discussed before, mechanical torque developed due to the interaction of two fluxes. Assuming constant field flux, torque became proportional to the flux produced around the current-carrying conductor. As flux proportional to the current, torque became proportional to the current and this is how we had derived $T_m = K_t I_a$. But when field flux is not constant, torque equation is,

$$T_m \propto \phi_f \phi_a \tag{14}$$

where ϕ_f is field flux and ϕ_a is armature flux produced around the conductor. So, it seems confusing that either speed increases or reduces when reduces field flux. Considering equation (14), when reduces field flux, torque developed also reduces and so speed. Thus, a reduction in field flux results reduction in speed.

Another fact is that back emf E_b also reduces with the reduction in field flux. Thus, it increases the armature current and so speed. According to this effect, a reduction in back emf results increase in speed. The answer to this controversial effect is simple. We need to understand which effect is predominant. One effect increases torque and the other reduces it as field flux reduces but the proportionality of both effects is different. Reduction in torque is less as compared to increase in torque when reduces field flux.

Thus, torque and speed increase with field flux reduction.

Equation of Force:

Let's derive the equation of force experienced by the current-carrying conductor.

Force produce at any instant is proportional to the current flowing through the conductor "I" and active length of conductor "l". Considering constant flux density "B", developed force,

$$F = BlI \text{ Newton} \tag{15}$$

where flux density $B = \frac{\phi}{A}$ weber/m².

Now consider Fig.15. Here force on the conductor is perpendicular to the magnetic field. But as we are interested in rotation of rotor assembly, only tangential component of force contributes. Considering the first rotor position of Fig.15 as reference θ=0, effective force or say useful force can be represented as

$$F_u = BlI\cos\theta \text{ N} \tag{16}$$

As the rotor rotates toward the magnetic neutral axis, a useful component of force reduces. Torque develop per conductor,

$$T_m = rF_u = rBlI\cos\theta \text{ N-m} \tag{17}$$

where 'r' is rotor radius. More numbers of current-carrying conductors are used to cover the complete rotor periphery to get maximum torque. The conductor may connect in series or parallel to form a winding ensuring the direction of current flows.

Shape and Equation of Back emf

Let's consider the motion of the conductor in a magnetic field as shown in Fig. 27(a). According to faraday's law, the magnitude of induces emf is given by,

$$e = \frac{d\phi}{dt} \tag{18}$$

Putting $\phi = BA$ in above equation becomes,

$$e = B\frac{dA}{dt} \tag{19}$$

Where 'A' is area covered by conductor travel. When conductor having active length 'l' travel by distance 'x', area 'A' becomes 'ldx' and generated emf,

$$e = B\frac{ldx}{dt} = Bl\frac{dx}{dt} \tag{20}$$

When conductor moves perpendicular to the magnetic field with velocity "v", an induced emf can be represented as,

$$e = Blv \tag{21}$$

Where "v" is the velocity of the conductor in m/s. But when the conductor moves parallel to the magnetic field as shown in Fig. 27(b), an induced emf is zero. Thus, considering "α" as an angle between direction of conductor motion and lines of magnetic field, generalized equation of induced emf per conductor can be written as

$$e = Blv\sin\alpha \tag{22}$$

Thus, when a rotor having numbers of conductors, rotates in the magnetic field with constant speed, an emf of sinusoidal nature induces. The total value of induced emf depends upon the series or parallel connection of conductors. As the commutator maintains the direction of current, emf induce at brush terminal remains unidirectional as shown in Fig. 28.

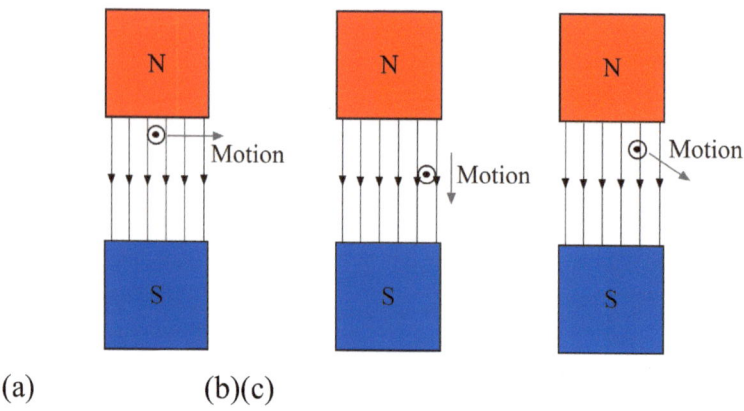

(a)   (b)   (c)

Fig. 27 Conductor moving in magnetic field

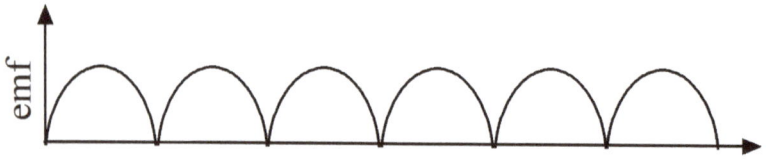

Fig. 28 Waveform of back emf

Equation (22) gives the instantaneous value of induced emf. Let's find the average value of induced emf in terms of number of conductor, numbers of pole, and motor speed.

Average Value of Back Emf

Let's consider 'n' as speed of motor in revolution per second (rps), 'P' as number of pole, and 'z' as numbers of conductors. Here 'ϕP' is the flux cut by the conductor per revolution, so average induced emf per conductor for speed 'n' can be represented as,

$$e = \phi P n \tag{23}$$

Consider 'N' as speed of motor in revolution per minute (rpm), then $n = N/60$. The above equation becomes,

$$e = \frac{\phi P N}{60} \tag{24}$$

Considering series-connected conductors, summation of induced emf for 'z' numbers of conductors is

$$e = \frac{z\phi P N}{60} \tag{25}$$

The above equation gives the average value of back emf induced in DC Motor.

Power Transfer:

Let's visualize the power flow from the electrical system to the mechanical system with the help of a mathematical equation. As explained in the block diagram of Fig.1 electric motor converts electrical energy into mechanical energy. Let's consider the electrical and mechanical power. Electrical power is the product of voltage and current whereas mechanical power is the product of torque and speed. Thus, electric motor can be represented as Fig. 29 where motor converts electrical power into mechanical power.

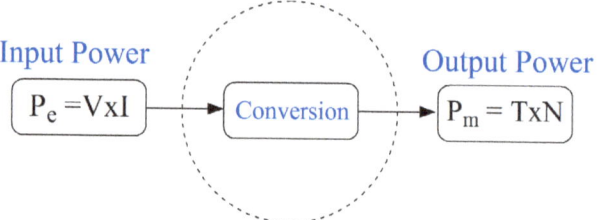

Fig. 29 Power flow in electric motor

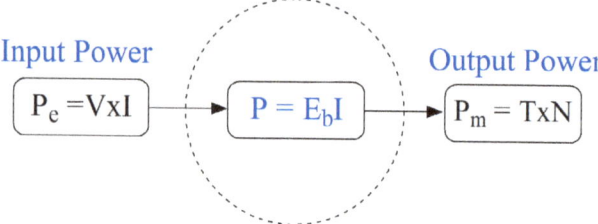

Fig. 30 Power conversion in electric motor

When the motor is supplied by voltage 'V', an armature current 'I' flows through supply (or battery) V_{dc}, R and back emf E_b. Here 'VI' is input power to the DC Motor and product '$E_b I$' is power opposite to supply power as direction of induced emf is opposite. Thus, electrical power "$E_b I$" is considered as equivalent mechanical output power. Power flow in electric motor can be represented as shown in Fig.30. Here "VxI" is electrical input power, product "TxN" is mechanical power output then "$E_b I$" is power transferred from electrical system to mechanical system. Here electrical and mechanical losses are neglected.

DC Series Motor

The main purpose of field winding is to generate a magnetic field. In DC shunt motor field winding is connected parallel to the supply voltage as represented in Fig. 22. Thus, current I_{sh} flows through field winding remains constant as

$$I_{sh} = \frac{V_{dc}}{R_{sh}} \tag{26}$$

And so magnetic field also remains constant. Here we need to understand a few more detail about electromagnetism. Recall the pattern of flux produces on the current-carrying coil shown in Fig. 9. Magnetic flux produce indeed depends upon the current flowing through the coil, but it also depends upon the number of turns. Higher numbers of turns result stronger magnetic field for the same current.

Thus, we have a choice to select numbers of turns and current to produce desired flux. As we are interested in minimum I^2R losses, we generally select higher numbers of turns and less current to produce desired magnetic field.

This is why we use field winding having higher resistance R_{sh}. Thus, field winding of DC shunt motor has higher resistance, thinner conductors, and large numbers of turns.

Here idea is to produce the same amount of flux with a higher current and fewer numbers of turns. Consider a field winding with thicker conductors, few numbers of turns and so very less resistance R_s. This winding can also be connected parallel to the armature winding to form a DC shunt motor. But as armature current is large enough, connecting this field winding in series produces desired amount of flux. Not only this, but by doing so we archives very interesting new characteristics of DC motor as very high armature current flows during starting.

This is because no back emf at zero speed and so full supply voltage appears across winding. And as winding resistance is very less, the initial current drawn by the winding is very high. As speed develops, back emf increases and current reduces. So, in the DC Series motor, field winding with few turns connected in series with the armature winding. Fig. 31 shows a circuit diagram of the DC series motor.

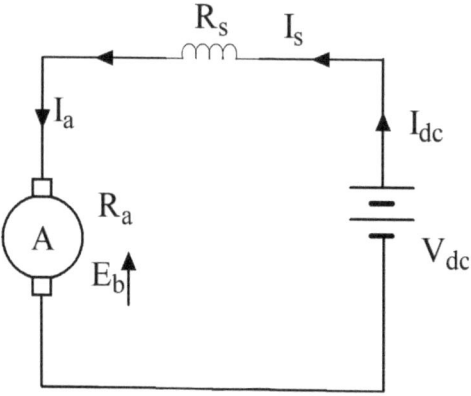

Fig. 31 Electrical circuit of DC Series Motor

Even if we assume series winding produces the same amount of magnetic field as produced by the shunt motor at rated current, the magnitude of field flux during starting will be very high in the case of DC series motor. Important characteristics of the DC series motor is that it produces very high starting torque. It made DC series motor more suitable for crane, hoist, and all traction applications which demands high starting torque.

Let's apply KVL and KCL to the circuit of the DC Series motor and derive the speed-torque equation. Current I_a flows through armature winding and field winding.

$$I_{dc} = I_a = I_s \tag{27}$$

Applying KVL, supply voltage,

$$V_{dc} = I_a R_a + I_a R_s + E_b \tag{28}$$

$$V_{dc} = I_a(R_a + R_a) + E_b \tag{29}$$

$$V_{dc} = I_{sh} R_{sh} \tag{30}$$

$$I_{dc} = I_a + I_{sh} \tag{31}$$

Here back emf is proportional to the motor speed as well as current which produces field flux. Thus,

$$E_b = K_b I_a N \tag{32}$$

Putting value of back emf in equation (29),

$$V_{dc} = I_a(R_a + R_s) + K_b I_a N \tag{33}$$

Let's simplify it for speed,

$$N = \frac{V_{dc} - I_a(R_a + R_a)}{K_b I_a} \tag{34}$$

$$N = \frac{V_{dc}}{K_b I_a} - \frac{(R_a + R_a)}{K_b} \tag{35}$$

Here torque is proportional to current that produces a magnetic field around armature conductor as well as current that produces main field flux. Thus, torque is proportional to the square of the current.

$$T_m = K_t I_a^2 \tag{36}$$

$$I_a = \sqrt{\frac{T_m}{K_t}} \tag{37}$$

$$N = \frac{V_{dc}}{K_b \sqrt{\frac{T_m}{K_t}}} - \frac{(R_a + R_s)}{K_b} \tag{38}$$

As discussed, the value of Ra and Rs is very small. Equation (38) reveals that at light load, the required value of torque developed by the motor Tm is very small and thus the speed of the motor is very high. The theoretical no-load speed of the motor is infinite whereas practically it is limited by friction losses and current limit. Even though the no-load speed and starting torque of the DC series motor is high enough to damage the motor parts. This is due to high mechanical and centrifugal forces produced by high starting current and high speed. This is why suggested to avoid starting a DC series motor with no-load condition.

Let's represent the equation (38) graphically. An increase in torque results in a speed reduction but it's not linear. Slight reduction in torque results in a large drop in speed as shown in speed-torque characteristics.

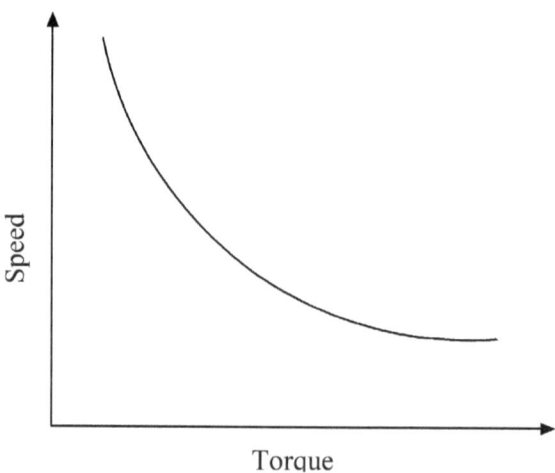

Fig. 32 Speed-Torque Characteristics of DC Series Motor

Shaft Torque:

Torque developed by the motor Tm is equal to the available torque at motor shaft for doing actual mechanical work when neglects frictions and windage losses. In actual motor, available torque at shaft is slightly less than torque developed by the motor T_m and it's called shaft torque T_s.

Applications & Limitations:

DC Motors are famous in variable speed applications as speed control methods are easy to implement. DC series motor is being used for decades for electric traction & industrial applications demand high starting torque. PMDC motors are widely used in electric toys and low-power domestic applications.

Major constraints that limit DC motor application are maintenance, cost, and size. Commutator and brush arrangement results issue of sparking when the current change the direction through commutator segments. This restricts the use of DC motors in a hazardous environment. Also, this arrangement is not suitable for dusty environments. It required regular maintenance to clean carbon dust from the commutator and may also require the replacement of brushes. Commutator also increases the cost of manufacturing and overall weight of the motor as compared to brushless motors. So nowadays it is common trend to use brushless motor for all industrial and domestic applications excepts few low power applications like electric toys and electrical hobby projects.

Appendix:

Faraday's Law:

It is about the generation of electricity with the help of a magnet and coil. It states that whenever flux linked with the coil changes an emf induces in coil. It was discovered by Michael Faraday in 1831. Mathematically it represents the magnitude of induce emf when flux-linked changes.

$$emf = \frac{d\phi}{dt}$$

Lenz's Law:

Lenz's law gives direction of induced emf. It states that emf induce in such direction so it opposes the quantity which is responsible for its production. Simply say's current produce due to change in flux linked produces another flux in opposite direction. It was discovered by Emili Lenz in 1834. Mathematically it is represented by a negative sign in the emf equation.

$$emf = -\frac{d\phi}{dt}$$

Screw Rule:

It is used to remember the direction of the circular magnetic field (flux) produces around the periphery of the current-carrying conductor. By knowing fact that we rotate the screw in the clockwise direction to insert the screw inward and anti-clockwise direction to take it out. When current flows inward, flux produces in the clockwise direction as explained in Fig. 8. This is analogs to the known fact that rotating screw clockwise results inward motion of screw.

Thus, by comparing the current direction with the direction of screw motion and the direction of the magnetic field with the direction of rotation of the screw, it's easy to remember the direction of magnetic field induces around periphery of current-carrying conductor.

Ohm's Law:

Ohm's law helps finding the direction of the current. It states that the current flowing through the conductor between two-point (resistive element) is directly proportional to the voltage across two-point. Mathematically represented by

$$I = \frac{V}{R}$$

where constant 'R' is known as resistance.

About Author:

Dr. Jignesh Makwana earned a doctorate in Electrical Engineering from the Indian Institute of Technology Roorkee in 2013. He works in the field of electric drives, power electronics & electric vehicle technology. He published many research papers & book chapters in reputed journals & conferences. He is the owner of "RhyMak Electronics" which is involved in manufacturing electric bicycles. He provides consultancy service to industries for product development & other technical supports. He delivers expert talk to national and international platforms in the area of their expertise. Besides all this, he is passionate about content creating and like books, chapters, technical notes, video courses etc. For more detail visit www.rhyni.com

www.ingramcontent.com/pod-product-compliance
Lightning Source LLC
Chambersburg PA
CBHW040326220526
45473CB00009B/2585